Awesome, Disgusting Science

GROSS SCIENCE OF POOP

Stephanie Bearce

BLACK RABBIT BOOKS

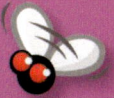

Hi Jinx is published by Black Rabbit Books
P.O. Box 227, Mankato, Minnesota, 56002.
www.blackrabbitbooks.com
Copyright © 2026 Black Rabbit Books

Alissa Thielges, editor; Jason Knudson, designer and photo researcher

All rights reserved. No part of this book may be reproduced in any form without written permission from the publisher.

Library of Congress Cataloging-in-Publication Data
Names: Bearce, Stephanie author
Title: Gross science of poop / by Stephanie Bearce.
Description: Mankato, MN : Black Rabbit Books, [2026] | Series: Awesome, disgusting science | Includes bibliographical references and index. | Audience: Ages 8-12 | Audience: Grades 4-6
Identifiers: LCCN 2025017501 (print) | LCCN 2025017502 (ebook) | ISBN 9781645824954 library binding | ISBN 9781645825012 paperback | ISBN 9781645825074 ebook
Subjects: LCSH: Feces—Juvenile literature
Classification: LCC QP159 .B43 2026 (print) | LCC QP159 (ebook) | DDC 612.3/6—dc23/eng/20250805
LC record available at https://lccn.loc.gov/2025017501

Printed in the United States of America.

Image Credits

Freepik/AI Image Creator, cover, 1, brgfx, cover, 1, 5, buchan, cover, 1, catalyststuff, cover, 1, freepik, cover, 1, 2–3, 5, 12, 19, GreenSkyStudio, 10, hittoon, 14, 23, macrovector, 11, 16, 17, nsit0108, cover, 1, OT Stok, 10, 11, rawpixel.com, cover, 1, shawlin, 19, studiogstock, cover, 1, 5, 6, 20, yojo, 9, 16, 17, 21; NASA/Karen Nyberg, 16; Shutterstock/Aleksey Matrenin, 13, AleMasche72, 10–11, Alex Coan, 15, AmolRoy, 6, Dilara Mammadova, 12–13, DM7, 15, Kamla S, 8–9, Krakenimages.com, 4, Kyle Waters, 9, mark higgins, 7, Memo Angeles, 20, Microgen, 19, Orange Dragon Studio, 12–13, Paul Vinten, 14–15, Phassa K, 7, PhotoPro Vault, 4, Suzanne Tucker, 4, SvedOliver, 11, TIGER KINGDOM, 8–9, VectorMine, 18, what is my name, 13, William Edge, 19, winfinity, 12; Wikimedia Commons/Los Perros pueden Cocinar, CC BY-SA 4.0, 16–17

Every effort has been made to contact copyright holders for material reproduced in this book. Any omissions will be rectified in subsequent printings if notice is given to the publisher.

CONTENTS

CHAPTER 1
The Scoop on Poop.....5

CHAPTER 2
The Experiments......6

CHAPTER 3
Get in on the Hi Jinx..20

Other Resources...........22

Chapter 1
THE SCOOP ON POOP

Brown and stinky. Everybody does it. POOP! If you eat, you poop. That's how the body works. It breaks down food into **nutrients**. The leftover waste goes out your butt.

Scientists love poop. It tells them about a person's or animal's health. But studying it can be downright disgusting!

A human poops about 25,000 pounds (11,340 kilograms) in their life. That's as heavy as three hippos!

Chapter 2: THE EXPERIMENTS

Poop Snack

Poop has trillions of **bacteria**. That's why it stinks. Scientists discovered baby koalas eat poop. It's a special poop from their mother called pap. It fills their gut with **microbes**. These break down the poisons in eucalyptus leaves. Koalas only eat this plant. Guess it's just poop and leaves on the menu. Dinner anyone?

A wombat's poop is cube shaped. It marks their home.

Poop Rockets

Humans poop on toilets. Then we flush it away. Not penguins! They shoot their poop out like a cannon. How far does it go? Researchers measured. They learned the poop rockets more than 4 feet (1.2 meters) away. This helps Adélie penguins keep their home clean without ever getting off their nest. How neat!

Flushing Fortune

All the poop that is flushed goes down to the **sewer**. People could be flushing treasure! Studies have found gold, silver, and platinum in sewer sludge. Small amounts of these metals are in our food and water. Experts think they could design a way to **mine** this poop. Then they would make millions!

There could be $13 million worth of metals in the city sewer of 1 million people!

11

Other biofuels come from plants and animal fats. They can fuel cars.

Poop Power

How would you like to ride a bus powered by poop? Stinking cool, right? Scientists are testing this with bacteria. These tiny organisms break down poop. They change waste into **biofuel**. One purple bacteria is already being used. It uses light to turn waste into hydrogen gas. Poop power may be the future!

Dino Dung

A poop fossil is called a coprolite. They look like rocks. These can come from dinosaurs. A pile can hold bits of plants or bone. This shows what dinosaurs ate for dinner. This is great news for scientists. They study hundreds of droppings. They see which dinosaurs chowed down on meat. Or they learn what plants a dino liked to eat.

Dino vomit can also become a fossil.

Astro Poop

How about a delicious serving of poop for lunch? That might be what future astronauts eat. Scientists are researching ways to turn poop into food. They use microbes. It would reduce the need to bring food from Earth. Deep-space exploring might be possible. Astronauts would never run out of food!

Poo that Heals

Feeling sick? A good dose of poop might be the **cure**. Researchers are doing poop **transplants**. They take poop from a healthy person and put it into a sick person. It goes into the intestine. Good bacteria grow and heal the stomach. This research could help with other diseases. It might even help cancer!

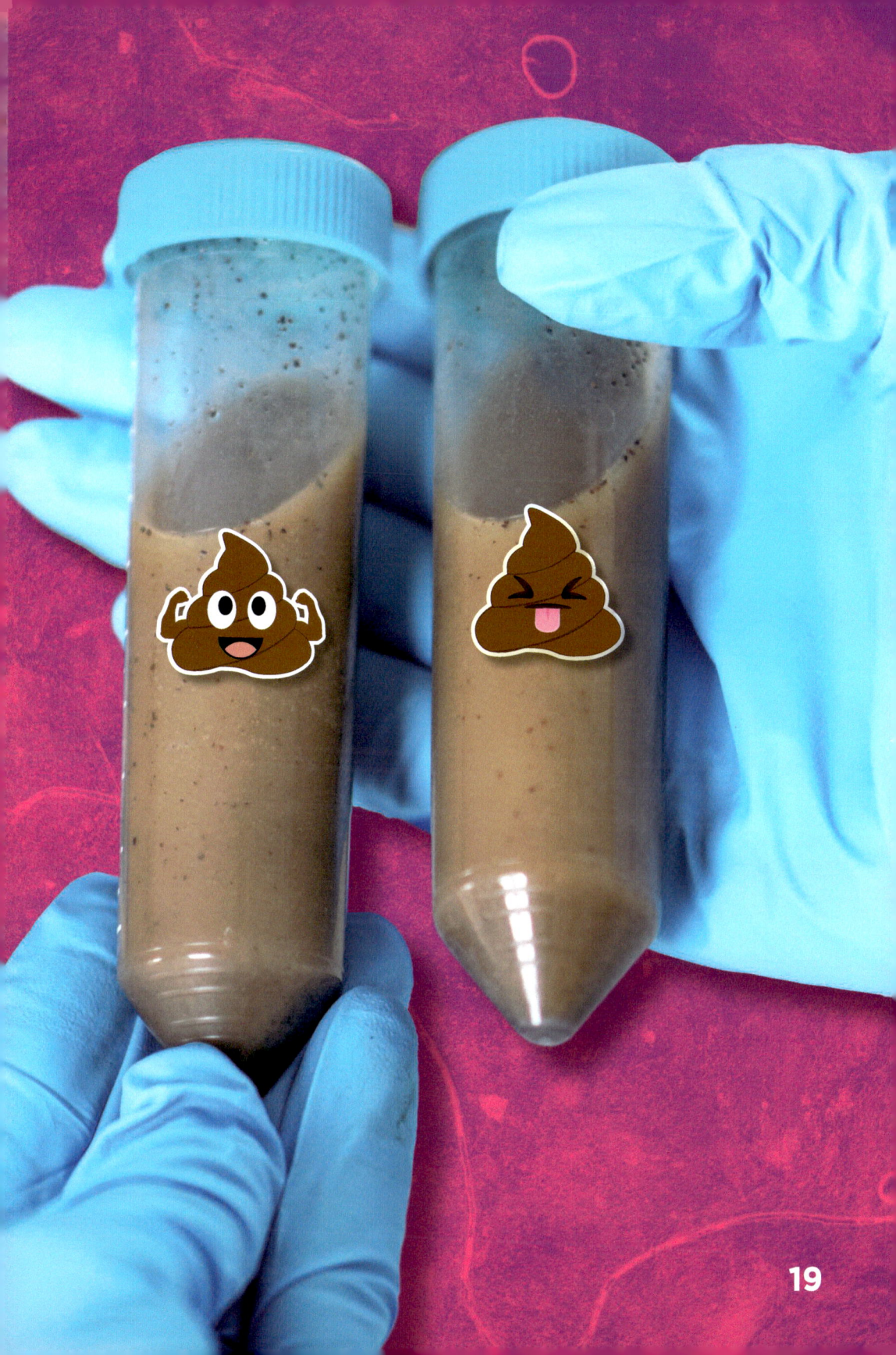

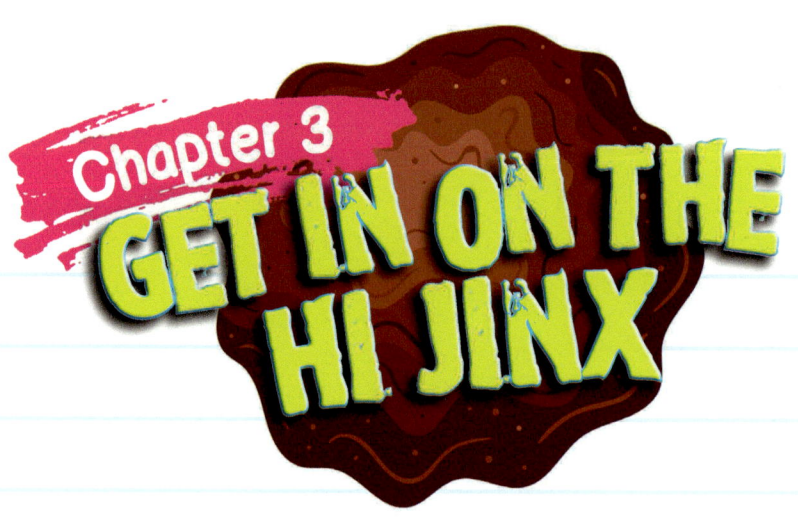

Chapter 3
GET IN ON THE HI JINX

Ever heard of a poopologist? It's a real job! These scientists are also called scatologists. They study poop of all kinds. They learn about health, diets, and even the environment. Want to become one? You'll need to study biology and ecology. Oh, and be okay with some serious stink!

Take It One Step More

1. If poop can power vehicles, what other things could it power?

2. How would you design a machine to mine gold from poop?

3. Why do you think studying animal poop is important for saving the planet?

GLOSSARY

bacteria (bak-TEER-ee-uh)—a very small living thing that often causes disease

biofuel (bahy-oh-fyoo-uhl)—a fuel that comes from living matter

cure (KYOOR)—something that stops a disease

microbe (MAHY-krohb)—a very small living thing that can only be seen with a microscope

mine (MAHYN)—to dig around to find something valuable

nutrient (NYOO-tree-ent)—a substance or ingredient a person or animal needs to be healthy

sewer (SOO-er)—an underground pipe that is used to carry off waste, like poop and pee

transplant (trans-PLANT)—a medical operation in which an organ or other part is moved from one individual to another

LEARN MORE

BOOKS

Bleckwehl, Mary E. *Disgusting Water and Sewer Jobs*. Mankato, MN: Black Rabbit Books, 2022.

Branzei, Sylvia. *Grossology: The Science of Really Gross Things!* New York: Grosset & Dunlap, 2025.

Leatherland, Noah. *Extreme Poop Facts*. Minneapolis: Bearport Publishing Company, 2025.

WEBSITES

Can Poop Cure an Infection?
tpt.pbslearningmedia.org/resource/nvgs-sci-poopcure/wgbh-nova-gross-science-can-poop-cure-an-infection/

Cool Jobs: Poop Investigators
www.snexplores.org/article/cool-jobs-poop-investigators

Poop: It's a Great Thing to Study
today.duke.edu/2020/01/poop-its-great-thing-study

INDEX

A
astronauts, 17

B
bacteria, 6, 13, 18

biofuels, 12, 13

D
dinosaurs, 14

F
food, 5, 10, 17

H
human poop, 5, 10, 17, 18

K
koalas, 6

M
microbes, 6, 17

P
penguins, 9

S
scatologists, 20

sewers, 10

T
transplants, 18